Science Conundrum

The Holy Grail of Science

To

The Cosmic Boundary

LAGARI A

CONTENTS

1 Abstract

The quest for the theory of everything begins when human curiosity leads us to observe and think, and this is also the birthplace of science. Scientists, philosophers, psychologists, and theologians have all attempted to come up with unified theories from their respective fields to explain everything, but these attempts have failed. As the quest continues, modern scientists have proposed this challenge.

I study these phenomena in order to find the technique of organizing the matter around us, whether it is by God or nature (whatever people may call it). If a central controlling consciousness exists, there must be common laws that are designed for the functioning of this universe. I am now disclosing the answer to the greatest question of humankind: the law behind everything, including the structure of life, and presenting a Grand Unified Theory that is acceptable to all fields of study.

I study the behavior of living beings and compare it to that of other materials. The interesting feature is that everything happening around us directly or indirectly influences us. My methods mainly include observations, simple experiments, and studying various research papers in this field.

The research is the culmination of the efforts of many renowned scientists like Newton and Einstein, and it overcomes the proposals of Gödel and a few others. I have finally understand that everything around us is connected to one another by means of energy waves. This theory satisfies the chaos and explains why some others have said that we cannot find the ultimate theory. My work also points out the certain properties of energy waves that maintain everything as it is, the singularity of all that we know and do not yet know.

I am exploring the channels of science, psychology, and philosophy to gain a complete scientific understanding about the world that we live in.

2   Towards Unification

In early times, during the discovery of fire, humans used their observational skills to understand the cause of forest fires. They found that fires were caused by the friction created when trees rubbed against each other during storms. After a long period of observation, they used the same rubbing technique with rocks and created the first artificial fire. This fire of curiosity ignited the minds of humans and led to the discoveries and experiences we have today.

As humans developed, they created theories, equations, derivations, and laws. These theories were born from observations of humans and their surroundings. In ancient times, and even still today, these observations are often put into statements. Some call these statements philosophy, while others call them theories. In the present scenario, philosophy and imagination are changing into theories and laws. Most often, recognized scientists use their imaginations and state their ideas with logical reasoning or mathematical equations. People often accept these ideas without question, assuming that educated elites would not state something that is untrue.

From ancient times, humans have wondered about the causes of everything around them. They have always known that the cause is not themselves, as the world existed long before their birth and will exist long after their death. The question then becomes one of finding the unified laws that govern everything. Humans finally found answers in external objects such as flowers, the sun, and the moon.

According to history, the concept of God evolved when humans began to worship external objects by giving them human features. When humans became sick, they found solutions in plants such as the Neem plant. They began to worship the Neem plant and this practice expanded to other objects, eventually leading to the creation of a number of gods. People taught future generations about their gods and their worship through stories. These imaginative stories gave gods human characteristics and features. Thus, the quest for the unification of everything led to the singularity of God.

When atheists arose, they began to question the existence of God and claimed that everything came from nothing, as the universe is electrically neutral. All the matter that we see is the positive energy of the universe, and the black space is the negative energy. They attempted to formulate a theory that would explain everything. This quest, which began with the origin of mankind, traveled through the greatest minds such as Isaac Newton and Albert

Einstein, and finally resulted in the conclusion that such unification is not possible.

Some scientists believe that the four fundamental forces were combined into one at the beginning of the universe (up to 10^{-43} seconds after the Big Bang). This has led to the emergence of a new theory called string theory. String theory is capable of incorporating each of the four fundamental forces in the universe. According to string theory, everything in the universe at its most microscopic level consists of vibrating strings or strands, each with a specific pattern of vibration. Due to the varied movement patterns of these strings, every particle has its own mass and force charge, created by a specific string pattern that is different from others. (That is to say, the electron is a string vibrating one way, while the quark is a string vibrating another way, and so on.) Many scientists believe that such a theory will provide a way to access 'the mind of God.'

What should a Grand Theory of Everything satisfy?

The theory must encompass not only science, especially physics, but also psychology, which deals with the unpredictable range of human emotions and behaviors. It must also account for both correct and incorrect experimental results. So far, scientists have focused only on unifying the four fundamental forces of physics, ignoring the need to include psychology. However, psychology has advanced considerably in recent years, and the unified theory must incorporate its insights, along with those of neuroscience, to explain the mind and its workings. The Grand Theory of Everything should account for all phenomena, from the tiniest movements to the creation and operation of complex systems, with absolute precision. Unlike other theories that only explain correct results, it should also encompass wrong, untrue, and unpredictable outcomes. It must answer why certain events occur in a specific way and why some matters are incalculable. Moreover, it must even be able to predict whether the reader will continue reading this text or not.

3 Just a few words

The methodology to find the Grand theory of everything is entirely different from other methodologies that we prefer today. The important thing is that as the Grand unified theory is the base of everything around us. So we have to observe the basic character and basic functionalities. As the theory is meant to everything in and around us, the only way of finding this is by observation and logical viva. The methods and final conclusions must be acceptable to everyone who thinks with logic and not with eyes.

Science Conundrum

What do you see from the top of a ten storied building in the center of the town which you are residing? Or simply what you see when you look outside? Obviously there were events taking place. People gathers, splits and don't even mind each other. It is like the twigs which are flowing the fast water current of a large river. All twigs are not from the same kind and are not from the same land. They all meet in the river. Some twigs move away from each other while others collide one another. This is a typical picture of a town. The vast river is the town and its fast current is the busyness of the town and twigs are the people in that town.

Genetic studies shows that we all belong to the same family which have its roots in the major continents and with the major civilizations existed in the past of time. But in the real world, we are just strangers. For a busy man he itself is a stranger. Imagine that you are walking through the crowded street nearby your locality. Suddenly a face flashes beside you. You turned back and that face also turned back to you and stared at you. But interestingly you both don't know each other. Then the quest rises, what makes you both to stare each other? Why you are attracted towards that person? And why this is not happening in the case of other people whom you met? The answer is in the hands of the universal law and engineering's which correlate everything.

4   First look to the Universe.

In order to understand the workings of the universe and its vast cosmic framework, we must consider the 3D coordinate system of space. Some physicists argue that taking more than three coordinates is acceptable, but for the purposes of our discussion, we will stick to the widely accepted 3D system. This coordinate system is made up of strings, which connect everything in space.

Figure. 1: Scan the QR to view the Artistic Illustration of Cognitor field

Figure 1 shows the cognitor field, which contains strings that resemble straight lines at their mean positions. These straight strings represent the "nothingness" of space, but in reality, there is no perfect absolute vacuum. The strings are actually interconnected and show deflections according to the presence of matter. These deflections create waves known as cognitor waves, which are unique for each and every object in the universe, from the smallest particles to the largest stars.

By studying and analyzing the cognitor waves in detail, we can distinguish and identify objects by comparing their waves. For example, the words in this book are composed of cognitor waves, and the book is a collection of a large number of cognitor waves. The collection of cognitor waves is called **Fluctus Samaharam**, and the entire universe is made up of an infinite collection of **Fluctus Samaharam**, which is known as **Ekathwam**. This is the final cognitor wave of the entire universe, and understanding it is a crucial step towards developing a unified theory.

Even the tiniest particle in the universe can consist of a large number of interconnected cognitor waves that provide information about its properties. These waves can add and subtract from each other to create a resultant cognitor wave, which is the basic wave that underlies all the forces in nature. These forces include gravitational, weak and strong nuclear, and electromagnetic forces, and they are either directly or indirectly influenced by cognitor waves.

The foundation of our understanding of forces lies in the work of Isaac Newton, who formulated the concept of force. Force is created when an object with mass is displaced with acceleration, and it is a product of mass, acceleration, and the sine of the angle between them. Force is also an attribute of physical action or movement and is sometimes referred to as strength or energy. The change in cognitor waves creates this force, and the force is simply an application or result of the change in cognitor wave.

If we ask what causes the cognitor wave to deflect from its mean position and create a force, we would eventually arrive at the Big Bang, which set the universe in motion and created the conditions that allowed cognitor waves to come into existence.

5 Big bang-Cognitor Wave.

The big bang itself was the result of a cognitor wave function, but we are not discussing what caused the big bang as it is the origin of everything, including the curiosity to question it. Just before the big bang, the entire universe was a tiny ball of energy that expanded rapidly, unfolding into the vastness that we see today. Right after the big bang, the universe was filled with an infinitely spread gas.

In 1982, a group of scientists including Stephen Hawking proposed an idea called "The Flaw of Imperfection." According to this idea, the uniformly attracted gas was affected by some imperfection, creating a change in its equilibrium and causing structures to form, defying gravity. While it may be difficult to believe in the idea of imperfection proposed by these cosmologists, what we consider to be imperfect may actually be perfect, and our inability to understand this leads us to think otherwise. Regardless, if this imperfection is indeed the reason for the formation of structures, there must be a cognitor wave that acts as the cause, and it should answer what caused the cognitor wave to create such imperfection.

The concept of creation is vastly different from historic notions. Consider the entire universe in its beginning as a small ball bearing. At each inflation, the ball bearing doubles in size but halves in mass, like a bouncing ball. [See Figure 3.]

The Big Bang was the result of a cognitor wave function, which unfolded from a tiny ball of energy into the vast universe we see today. In 1982, Stephan Hawking and a group of scientists proposed the idea of "The Luck of Imperfection" to explain the formation of the universe's first structures. They suggested that imperfections in the uniformly attracted gas, caused by a change in equilibrium due to gravity, led to the creation of these structures. However, it's difficult to see imperfections as anything but perfect, since our inability to understand them is the only flaw. Regardless, if imperfection did play a role, there must have been a cognitor wave that caused it, and its origins remain a mystery.

The creation of the universe involved a process of bouncing, inflating, and distributing mass and energy through cognitor waves. The original ball of energy bounced and doubled a million times, distributing its mass and energy to vast copies of itself. Over time, the universe evolved from singularity to entropy, and collisions between re-bounced balls created new masses. The fluctuations in the cognitor wave's speed increased the collision rate, and the universe spread out into space in a small amount of time. The second law of thermodynamics was a key factor in this creation process.

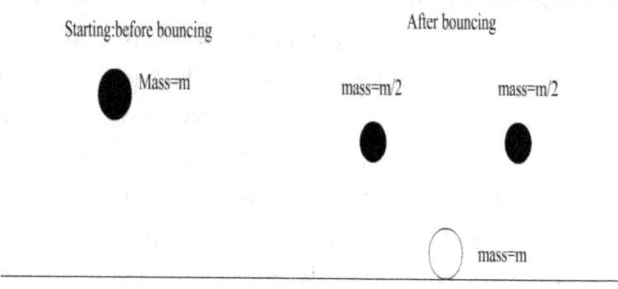

Figure 2 Illustration of the creation with inflation.

To illustrate the vastness of the universe, imagine a small universe with singularity energy of 10000 kJ and

```
Mass distributed into :  5.555555E-13 Kg



FINAL STRUCTURE OF UNIVERSE AFTER 100 EVENTS AND SINGULARITY ENERGY OF 1E+07
JOULES
1) Total mass of space :  1.111111E-10 Kg
2) Total energy of space :  1E+07 Joules
3) Total No of quanta :  200
4) Energy per quanta :  50000 Joules
5) Mass per quanta :  5.555555E-13 Kg



-----END OF RESULTS-----
Press any key to continue
```

involving 100 events (inflations). The resulting universe would have an energy of 1×10^7 Joules, distributed to 200 quanta with 50000 Joules per quanta. When this energy was converted to mass, it was valued at 5.555555×10^{-13} kilograms. A computer-generated output (Fig. 4) can help visualize this process.

Figure. 3 Computer generated result

The cognitor wave's specific ways of interaction are the ultimate controller of the universe's functionality, as demonstrated by the "game of life" devised by mathematician John Horton Conway in 1970. Understanding the grand engineering of the cognitor wave in creation requires us to consider these interactions carefully.

6 Engineering of Cosmic Masterpiece

The origin of the universe has been a topic of debate for centuries. In medieval philosophy, one of the central questions was whether the universe was finite or infinite. Aristotle believed that space was finite but time was infinite. However, this concept did not sit well with Islamic, Jewish, and Christian philosophers who felt that the Aristotelian concept of eternity did not integrate well with the Abrahamic view of creation.

One of the most widely accepted theories about the origin of the universe is the Big Bang theory, which was proposed by Belgian priest Georges Lemaitre. Lemaitre, who was not only a priest but also an astronomer and professor of physics at the Catholic University of Leuven, first proposed the theory of the expansion of the universe. His theory is often misattributed to Edwin Hubble, who was the first to derive Hubble's law and Hubble's constant. However, Lemaitre published these two years before Hubble's paper in 1927. Lemaitre's "hypothesis of the primeval atom," also known as the Cosmic Egg, is now widely known as the Big Bang theory of the origin of the universe.

The Big Bang theory is an attempt to explain what happened at the very beginning of our universe. According to the theory, everything popped up in a fraction of a second, starting from nothing at all. At its simplest, the Big Bang theory describes the beginning of the universe as a singularity, which then began inflating over 13.8 billion years to become the universe we see today. While some might describe the birth of the universe as an explosion, it is more accurate to describe it as an expansion.

In 1610, German mathematician Johannes Kepler argued for an infinite universe based on the concept of a dark sky. However, it wasn't until seventy-seven years later that Isaac Newton described the large-scale motion throughout the universe.

In the early 1910s, Vesto Slipher used spectroscopy to investigate the rotation periods of planets and the composition of planetary atmospheres. He was the first to observe the radial velocities of galaxies, while Carl Wirtz observed a systematic redshift of galaxies. These observations later supported the idea that the universe is expanding.

In the same decade, Einstein's theory of general relativity, based on his field equations, suggested that the universe was described by a metric tensor that was either expanding or shrinking. According to the standard model, the universe came into existence from a singularity billions of years ago. Singularities are thought to exist at the core of black holes, which are areas of infinite gravity where all the mass gets crushed and packed closely together to form infinite density.

The universe is said to have begun from a singularity that was incredibly hot and small. It then began to expand and cool, eventually forming everything that we see today, including ourselves. It's important to note that the singularity didn't appear in space; rather, space began inside the singularity. Prior to the singularity, there was nothing - no space, time, or energy. Simply put, the entire universe began from nothing.

7 Grand Unified Theory of Everything

Have you ever wondered about the mysteries of the universe and how everything works? From the tiniest subatomic particles to the largest galaxies, there are still so many questions that remain unanswered. Many brilliant minds throughout history have tried to make sense of it all, and one of the most ambitious attempts is the string theory.

Picture a stage filled with strings of all different lengths and widths, vibrating in a harmonious symphony. That's what the world would look like according to string theory. In this theory, the point particles of particle physics, like quarks and electrons, are replaced by tiny one-dimensional strings that vibrate at different frequencies. And these strings have something special - they possess extra dimensions beyond the three spatial dimensions and one dimension of time that we are familiar with.

As these strings vibrate and move, they can create something like a membrane called a Brane, which is a way of describing the 11 dimensions of string theory. That's right, 11 dimensions! It may be hard to wrap your head around, but this theory suggests that there are hidden dimensions of space that we can't see or interact with directly.

String theory attempts to explain all elementary particles based on the quantum states of these strings. This includes even gravity, which is usually explained by Einstein's theory of general relativity. And in fact, string theory combines general relativity with quantum mechanics, which is a major challenge that has eluded physicists for decades.

But here's the catch - string theory is not just one theory, but a family of theories that have evolved over time. The original string theory was actually five different theories that were later found to be different aspects of a single theory called M-theory.

The cognitor theory takes things a step further by proposing a complete theory that unifies all irregularities and anomalies, making it a true theory of everything. It's not just about predicting outcomes, but also about explaining everything from the fundamentals of energy and force to the complexities of psychology and psychiatry.

To do this, cognitor theory combines Newton's equations with Einstein's famous equation $E=mc^2$, which connects energy and mass. Einstein's work was a breakthrough in science because he showed that energy and mass are two sides of the same coin. And now, cognitor theory builds on that foundation to provide a more comprehensive understanding of the universe.

So, while the mysteries of the universe may still be unsolved, string theory and cognitor theory offer fascinating possibilities for a deeper understanding of everything around us.

To have a theory of everything, we need to explain the fundamentals of energy and force. The cognitor theory attempts to do just that and unify all aspects of science and psychology. Its applications will be discussed in later chapters.

7.1   Force

Let us again consider the explanation of Big bang. Scientists claim that the expansion is caused due to pressure difference and some other factors. Then we want to ask that what is pressure? The probable answer is force acting per unit area. Then what is force? By Newton's definition;

"Force: Anything that changes or tends to change the state".

It is a good explanation. But what is that anything and what is the basic anything in the definition. Anything is the change: change in surrounding: concentration change of energy.

Then by substituting the word change in the definition of force by Newton we get: -

"Force: Change is that changes or tends to change the state".

This is the action-reaction chain hence formed;

Action↔ Change in surrounding↔ Change in cognitor waves↔ Change in energy

Force creates action. Action can be classified into two;

 ➢ Movement base.
 ➢ Non-movement base.

Movement base.

     Movement base actions are those actions that can be seen (its effects) with naked eye and can calculate with concrete techniques identically.

Non-movement base.

     Non-movement base are those that can't be seen (as a change which can measure by any instrument accurately) but can feel sometimes.

E.g. Love, Attraction, Meditation etc.

This is similar to the chaotic condition for the change sequence.

7.2   Energy

Everything around us including mass is a form of energy. We also know about different forms of energy in our daily life like light, chemical, electrical, potential, kinetic etc. Then it is possible to have a unified energy or a basic energy that cause the creation of other forms. For that we have to move backwards, backwards even before our creation to the Big bang.

The expansion of big bang is considered as the father of everything that we see including us. The whole universe is getting created at the time of Big bang and even the time itself is created by Big bang. Here we solve the unanswered question by the expansion.

During an expanding explosion (consider Big bang), energy is get emitted and after the emission it soon changes into different forms. These changed forms of energy are the ones that we experience today around us. The energy that arise at the time of expanding explosion i.e. at the time of origin of universe.

Here a question arises, then why we are not known about such energy? The answer is simple that is our understanding is limited to the time of birth of "Time" itself. The time have its origin when the light origins. Therefore it is clear that the 'basic energy' called 'Ekaoorjam' is converted or transformed to other energy forms and this transformed energy is resulted in the formation of heat during the Big bang expansion.

We now know where the first energy formed, but what is this "ENERGY"?

The probable answers include the opinion that energy has various definitions. The main two definitions are as follows:

- ➢ Something that have the ability to cause change.
- ➢ The strength and vitality required for sustained physical and mental activity.

As energy is something that have the ability to cause change and then why don't we have the unified energy? It is because from the first explanation / definition of energy, it is clear that energy is something that cause change. After causing the change, that change leads to cause another one. Finally all these chain of changes results in the formation of numerous forms of energy that we experience today.

Relation between force and energy.

"Force: Change is that changes or tends to change the state. -1

Energy: Something that have the ability to cause change."   *-2*

From 1 and 2 we get;

Energy is the cause of force and both appear in pairs.

Therefore;

The functionality of everything around us is summoned to a single statement that:

Each and everything is a function and functional unit of energy.

This is the law of everything. The theory and law of everything is just this simple statement.

7.3   Work

Work is said to be done when some force displaces a body. In physics work is said to be zero when either force or displacement is zero or force and displacement are perpendicular to each other. Simply in physics if you carry some load in your head and walk to your home is not considered as a work nevertheless of your effort. This is because force acting due to gravity is downwards whereas your displacement of load is parallel to ground and is perpendicular to force of gravity.

$$W = FS\,Sinx$$

Force is executed by movement of cognitor waves. Simply work is taken place when there is a displacement of cognitor waves (wave interaction) i.e. energy change.

Hence,

Change in state of energy is work.

7.4   Derivatives

These statements further expands to a set of laws which is their functional properties.

They are:

7.4.1   Each and every thing in the universe generates energy.

We live in an electronic wonderland, surrounded by a vast array of electronic devices, each humming with their own unique energy. And what is this energy? It's electromagnetic radiation! Think of it like a sort of cosmic symphony, where every electronic device is producing its own unique set of notes in the form of waves. But these waves are more than just pretty to look at - they're actually incredibly useful!

In the world of cybersecurity, these waves are used to decode data that's being transmitted between electronic devices. It's a process known as TEMPEST Attacks, and it's like something straight out of a spy thriller. Imagine a skilled engineer hunched over a computer, analyzing the faint electromagnetic signals being emitted from a nearby ATM, and using that information to extract a person's personal data. It's like something straight out of a James Bond movie!

But it's not just electronic devices that emit energy - our bodies do too! According to spiritual healers and parapsychologists, there's an energy flow in and out of our bodies, and they call it our aura. It's a sort of ethereal glow that surrounds us and is attributed to divine forces. But this isn't just a figment of their imagination - they actually have a way to prove it!

Enter Kirlian photography, a technique discovered by a Russian electrical engineer named Semyon Kirlian in 1939. This technique uses high-voltage electricity to photograph the coronal discharge (or aura) around objects. And while it's not widely accepted in mainstream science, it's a favorite tool of spiritual healers and alternative medicine practitioners who use it to analyze and diagnose the aura of their patients.

But what does all of this energy mean? Well, it turns out that the waves produced by electronic devices and our bodies can tell us a lot about the world around us. By studying these waves, we can determine the state, nature, and condition of objects, as well as their ingredients and other important information. And the interaction between these waves even creates a force - the gravitational force - which is what keeps us all from floating off into space! So the next time you turn on your electronic device, or catch a glimpse of someone's aura, remember that you're surrounded by energy in all its beautiful and mysterious forms.

7.4.2   *The greater energy (impulsive force) will win (due to greater momentum).*
He who has the ability would be the leader (Malayalam proverb)

7.4.3   *There is only one thing called energy (for our convenience we termed it as positive negative and neutral)*

7.4.4   *There is an emitter and receiver between the energies.*

The energy radiated from a body get absorbed by nearby bodies or get attenuated in space.

7.4.5   The receiving energy from a body depends upon the nature of the receiver.

Wave mechanics takes place. Sometimes waves get added or subtracted each other resulting in new wave or nothing. The radiating field around the body decides whether an outer wave can interfere the body.

7.4.6   Frequency of energy varies according to condition and nature of body.

The frequency of radiation is influenced by the state of the body.

7.4.7   The flow back of energy varies according to the condition (Health) of emitter, receiver or even both.

7.5   Health of the system.

The state of body depends on certain factors that we can call as Health of the body or system.

Health includes the following;

7.5.1   State of matter.

State of matter includes all known states and unknown states of matter. A solid is defined as a form of matter which poses rigidity. The constituent particles in the solid oscillate in their mean positions. A crystal is made up of atoms arranged in specific order. The order of arrangement tells what crystal it is. A regular 3-D arrangement of particles in space is called crystal lattice. The smallest portion of the lattice is termed as a unit cell (atom). The arrangement is different in liquid, gas and so on.

7.5.2   Color of the body.

It is a result of electromagnetic waves and how brain perceive it. The decoding range of waves of color varies from organism to organism.

7.5.3   P-factors

These are factors that is applicable to living bodies that have ability to think and decide, example: humans. This is sub divided into

7.5.4   Psychology

7.5.5   Data analyzation of brain

7.5.6   Response to both mental and physical stimuli.

7.5.7   Motion of the system

If we are considering a mechanical system (any system can be considered). Then the internal motion executing by the system also influences the cognitor wave radiated as part the system. In case of human body the motion of ions, blood, vitamins, food, air etc. are accounted.

7.5.8   X-factors

These are all other factors which effect the radiation of energy from a system.

8   AURA, ENERGY AND SCIENCE

You may hear of people healing others through their looks and certain hand positions (Reiki). Sometimes they predict your future, past and all about you. How these 'black magic' possible? Is it true? Science is here to reveal it for you.

Reiki and other healing and predicting techniques are based on energy, namely Aura (bio energy). Aura is a seven level bio energy field that gets activates when the body stabilizes the energy inside itself by concentrating. Science proves that the body which is not concentrated actually distributing its bio energy to the surroundings. The bio energy is the energy that is stabilized in our body. The concentration help to stabilize it to a point i.e. to our body itself and hence the wasting of energy to the outside is not happening and hence the energy rise as it is not distributing to all parts. The first stage of concentration (meditation) involves stabilization of energy for the formation of Alpha state followed by Beta and Gamma in the following stages.

The capturing of these waves which is got emitted from the body help us to study about that the behavior and every aspects of that particular body. These normal waves move like a light beam whereas a wave from the concentrated body moves like a laser beam. Therefore concentration can act as a better force to create a notable change.

9 NEUROLOGY AND PSYCHOLOGY

As per a recent research studies each thought is associated with each breaths. The concentration involves concentrating our thoughts to one and later in a state of zero thought. (Zero thought doesn't mean that the body is not having breath. Breathing is an easiest way to get merge into single thought.) It can be done by concentrating in our breath. This clinically proven and practiced by millions around the world.

As we concentrate in one thought and later merge into a state of nothingness. This is where science plays. The feeling of nothingness is due to the role played by the neurons in the brain. The brain opens its neural channels. In the case of a normal being neural channels are active for only a certain areas, the areas where he is concentrating more in his lifetime. I.e. the active neural channels of a math teacher is entirely different from that of the active neural channels possessed by a musician. But the concentration of energy to oneness involves opening all the neural channels and making the whole channels in our brain active. By concentrating we are allowing the energy to activate the neurons even though we are not utilizing. This increases the efficiency of the brain. Therefore it is suggested to meditate for many health problems and even the 'cool off' time before exams.

10 MIRROR NEURONS AND REALITY

We all see movies, hear music and have other entertainments. Do you know why we are entertained when we hear a favorite music or attend an event? The entertainment itself is a science. It is not depends on how the movie projector or the I-pad works. Nothing becomes entertainment until we enjoy it. Then how we are enjoying? Enjoyment is a feeling that we are also the part of what we are indulging. Seems to be philosophical right? But science is behind this philosophy. Our brain possess certain neurons called "Mirror Neurons". These are the channels which help us to access others mind and enjoy what others have. Consider.

A man is playing game in his computer. We can see that the man is behaving as if he is inside the game and the game seems to be calling the man to play it more. We feel he is also a part of the game. How it is possible? Both game and person are entirely different ones. But he is in the game. This is where the mirror neurons play their role. It is brain which receives and processes the images and sounds from every part of the body and the brain is network of millions of neural connections. The brain try to create the same neural connections that is responsible for the enjoyment and these is done by the neurons by creating the same connections that the creators intend to connect and to feel. If the man's brain is not working properly the ability to enjoy or experience is not fully get into action.

One of the reasons is neuroplasticity. Recent neurological research has confirmed the existence of empathetic mirror neurons. When we experience an emotion or perform an action specific neurons fire, but when we observe someone else performing this action or when we imagine it, may of the same neuron fire again as if we were performing the action ourselves. These empathy neurons connect us to other people, allowing us to feel what others feel.

And since these neurons respond to our imagination, we can experience emotional feedback from them as it came from someone else. The system allows us to self-reflect.

Sometimes I used to have some romantic dreams. Among one I saw me kissing a girl. The interesting fact is that I suddenly wake up as kissing is just unacceptable for me. But even after when I wake up I feel the taste of her lips in my tongue. When I lick my lips I can taste her lips. How is that possible? Even after a month I feel it whenever I lick my lips.

We have developed technology that enables us to read human mind. The computer is programmed to function as our mind says. The computer here analyses brain waves to decide what the need is and what action to be performed. These brain waves also constitute its cognitor wave. Like mirror neurons we can recreate the incidents of ones memory to others by simply recreating cognitor waves.

11 In a nutshell

11.1 Everything is made of energy.

11.2 The basic energy caused the basic force.

$$\because \quad {dE}/{dx} = F$$

I.e. Force is change in energy with unit displacement.

11.3 Everything is connected to everything else.

11.4 The connection is carried out by cognitor waves.

11.5 Cognitor waves varies each other and act as a cosmic identity for each body or system.

11.6 Even the tiniest particle or disturbance contains more or less several cognitor waves.

11.7 Creation and its expansion is a result of cognitor wave action.

11.8 The basic energy and basic force constitute the all forms of energy and force that we see today.

11.9 The workings of energy waves are responsible for the events that are get created around us.

11.10 It controls the physical and psychological world.

11.11 It proves everything despite of good, bad, wrong etc.

11.12 It satisfies every set of answers and future discoveries.

11.13 Cognitor waves of living body can be used of studies and varies test and even in curing diseases.

11.14 The grand unified equation.

$$E_f = \sum_{i=1}^{n} f(e)$$

or

$$E_f = \sum_{i=1}^{n} e$$

Where E_f is the final event, f (e) is the factors or function of that event that affects the event E_f,
The final event E_f or final result is the sum of all the events or factors that affect the event.

12

APPLICATIONS OF COGNITOR FIELD THEORY OF EVERYTHING.

APPLICATION: 1

12.1 REALITINESS OF REALITY:
existence of universal boundary

The concept of reality is a fundamental aspect of our existence. It refers to anything and everything that we experience and consider to be real. It may be an incident or any other matter that we perceive through our senses. However, our perception of reality can be influenced by various factors such as media and scripted events. For instance, reality shows are designed to entertain us, but we only see the reality that is presented to us on a 2D screen, which is often scripted and manipulated. Our understanding of reality is limited to what is shown to us, and we are often unaware of what lies beyond the screen.

According to the Vedas and Naadi Shastra, reality is the time of existence. They measure existence as a time called Naadi, which represents the time between one complete exhale and before the next inhale. Therefore, the reality exists only for a period when we are not inhaling air. This leads to the conclusion that there are about 14 realities in one minute.

Reality is a present tense word, which means it refers to the moment of existence. The present tense has a short lifespan, while the past has an increasing lifespan, and the future is only a curiosity about the next present. Our well-defined senses enable us to perceive reality, and vision is the primary sense that we rely on. However, for us to see, we need light, which is the major ingredient that enables us to feel and see reality.

The phenomenon of light plays a significant role in how we perceive reality. We can represent the past, present, and future by a light cone, and every event in the present seems to take place because of our feelings. The feeling of experience is created by light and is sensed by the brain, which then interprets it as reality. The Einstein-Murkowski model of space-time explains this idea through a visual representation where the point P represents the past and the point F represents the future. This concept is like ripples in water, where the present moment is represented by the ripples and the past and future are the calm water around it.

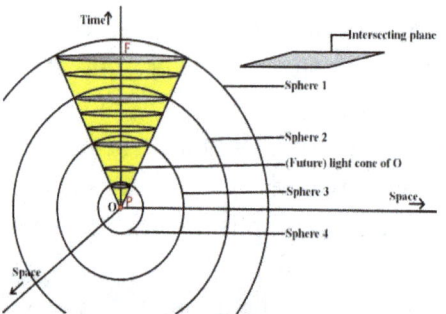

Figure. 4 Light cone represented by ripples in water.

In the above *figure.6* the P[present] represents the time when a stone hits the water body. The ripples firm and create a cone. The point F represents the future.

The point Q in the *figure.5* shows the area which is not present inside the light cone and thereby not exists. This is because the points outside the light cone are not calculated as light is not going through that part. Here is a case. Consider that you are searching for your pet in darkness inside your room with the help of a torch. The light from the torch moves same as like the light cone. In the darkness of the room, you are pointing the torch to many points. You can only see the things inside the light cone of reality and according to that, those things that are inside the light cone of your torch only exists and the rest didn't. That means as the torch is in your hand and the light is not focusing at you.

Therefore, you also didn't exist in the frame of reality.

Figure. 5 Area of reality

The only way to avoid the inexistence due to the in appearance in the field of light cone is that to spread the light to every side in a way that it falls on all bodies which is need to be in existence, or in the frame of reality.

Let's do that

Figure. 6 Merging of two realities.

In the above *figure.8*, the light fall on everything and it falls back to the source itself by its property of absorption and re-emission, to make the source also in the frame of reality. The points R and S show the intermixing of individual light cones.

Now it's time to rewind the *figure. 6* which shows the ripples model of light cone. Ripples are the waves created in the water. Like that, light is an electromagnetic ray which possess both particle and wave nature. We are here taking the wave nature of light as ripples in water. Is water simply consisting of a single wave?

For understanding the idea clearly we have to imagine a pond during rain. Each raindrop makes each light cones in the water [as per *figure. 6*].The rain doesn't consist a single drop, it consists of a million and millions of drops in which a few thousands will fall on the pond depending on the size of your pond. Consider two drops A and B falls in the nearest points in the pond [*figure. 9*].The light cones created by each drops gets mixed and the wave get added and cancelled.

Figure. 7 Two drops hits in a water body

The points P, Q, R, S and V are the points of coagulation of waves which is created by the drops

A&B and called false points. These false points create a coagulation area of PQRS, which represent false reality. Due to the coagulation, the reality which is taken by each drops A&B get cancelled or added and gives a new reality. The new reality may be a sum of reality of A and B or not a reality due to cancellation of waves of A and B.

Now we can go deep into the above phenomena with replacing the water and drops by sun.

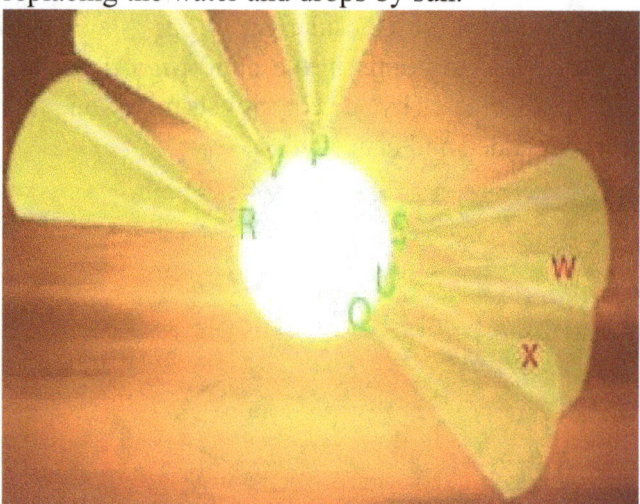

Figure. 8 light cones from sun (illustration)

Sun is a star with spherical and circular in shape. Mathematically a circle is made up of infinite number of points. This infinite number of points creates infinite number of light cones. Thus it is enabling us to see the space around the sun, including us in the earth. But when infinite number of light cones originate, there exists infinite number of false points and false realities. Therefore the reality which we experience is the culmination of false realities. The realitilessness of reality is formed or happened in the case of a distinct star. In fig 10, it illustrates the sun and its few light cones among the millions and millions of light cones. The points W and X represents the area of False realities. As the light cone develops, more and more false point's got created in the universe. Consider a star which is 100 light years far away from earth. If the star dies in 2013, we will only know about that in 2113, as the last light from the star reaches earth by travelling a distance of 9.46×10^{17} in which is 9.46×10^{14} miles! And $2.353702229 \times 10^{11}$ time the perimeter of earth. Science also explains the big bang creation by the means of light cone. The *figure. 11* shows the light cone representation of big bang.

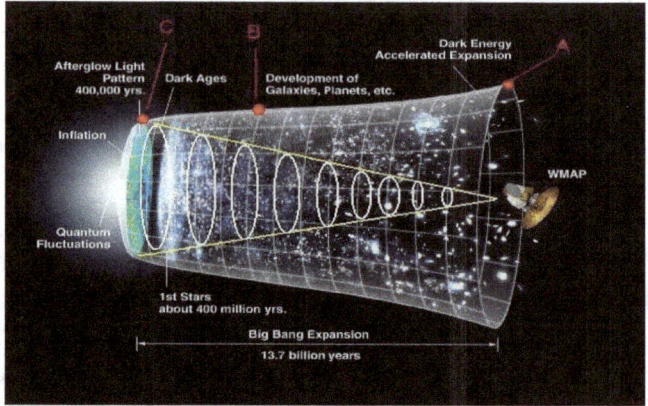

Figure. 9 Light cone representation of Big Bang

According to the fig 1.7, a question arises here is that the light have its birth long after the big bang. And when we use this light cone as a reference of study we will not be able to get the knowledge about the real big bang and calculate its date of birth.

Without intending to question the light cone concept of science as science only exists by means of concepts and theories which can satisfy the general and majority. Science also always put forward a great twist i.e., exception in everything.

12.1.1 BRIEF-BIOGRAPHY OF LIGHT

Now let's discuss the light cone in the true manner of science and in a different manner of logic. We begin with the birth of light. When sufficient energy is supplied to an atom, the outermost electron of the atom jumps to the exited state. This time, as the potential energy of the electron that jumped to the new orbit increases and which lead instability in that shell. In order to retain the stability, the electron jumps back to the ground state from its exited state. The energy which the electron absorbed when it jumped to the exited state from ground state gets liberated during its journey from exited state to ground state. This liberated energy is in the form of energy packets or quanta. In the case of light, the quantum of energy is called photon. Thus light is produced.

Now we know about the production of light by the jumping of electron. Light consists of a lot of frequencies and only the visible spectrum of light is from 400 to 750 nanometers. This spectrum of electromagnetic radiation was obtained by the hydrogen spectrum experiment. The *figure. 12* shows the various types of electromagnetic radiations which differ from one another in wavelength and frequencies.

From the *figure.12* it is clear that the visible spectrum of light is too short when compared to the whole spectrum of light.

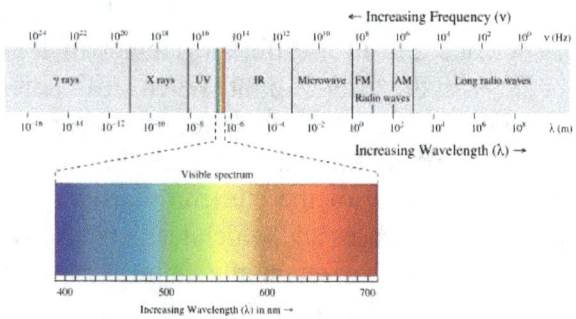

Figure. 10 EM wave spectrum

12.1.2 STRING WAVES

All waves possess a velocity, simply wave velocity. It depends on the crusts and troughs it creates. In each oscillation the wavelength and amplitude of the wave decreases. For getting this better, consider a string of guitar. When the player touches the string of guitar a wave is created and thus produces sound. Within a short while, the sound ends and the string comes to rest. At this moment the string have its crusts and troughs in line of origin. This is because of the decreasing amplitude of the string due to oscillation. The *figure. 13* illustrates the dissolving of waves.

The decreasing of amplitudes of strings is represented in string cone. All strings posses' string cones. In the above string cone *figure. 13*, the point

A is formed first and as time goes on point B is formed.

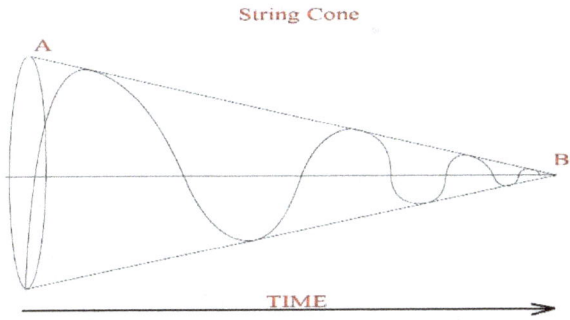

Figure. 11 String cone

String cone is applicable to all waves whereas light cone is applied to path of light from the origin. As light is an electromagnetic wave, and posse's particle and wave nature. Thus the light possess both light cone and string cone. The light cone shows the paths which light sweeps whereas the string cone shows the nature of light waves. The *figure. 14* below represents the light cone and string cone of light with respect to time.

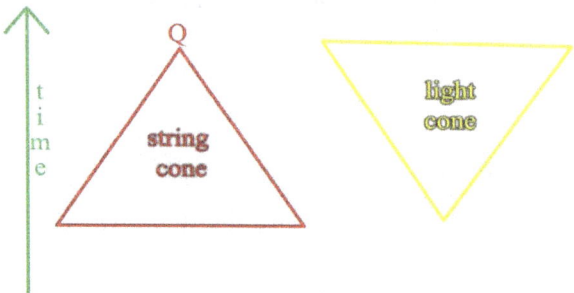

Figure. 13 Light and string cone with respect to time

The point Q in the *figure. 14* shows the start of waves

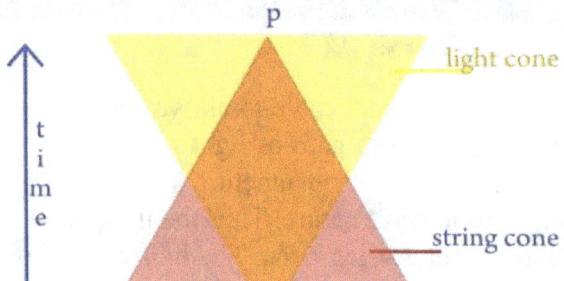

Figure. 12 Superimposing light and string cones

that from that point string doesn't have a wave nature (as per our eyes). Therefore at Q there is no wave or the waves emerge to the surrounding space. At that point the wave loses its ability to continue its journey as a wave.

Now we can insert the string cone in the light cone as per the *figure. 15*.

As per the *figure. 15*, we can understand that when light cone expands the wave nature [amplitude decreases].At point P when the light cone expands to its most, the light waves emerges to the space. Thus the lights have its death at the point P. The decreasing of amplitude of wave of light is to keep the velocity of light a constant. If the amplitude is high and constant, then on moving, the velocity decreases due to its motion. So waves try to decrease their amplitudes to keep their velocity constant [in the case of light wave]. This simply sounds that the light has its death after its long journey.

12.1.3 Universal Boundary

Therefore, the realities also get dissolved in the pass of time. Now we can take the universe, the land of varied realities into our study. The universe is an endless bordered space with stars and other celestial bodies. According to string cone, the wave of light only can travel to a specific distance. If the universes have a boundary outside that distance, then we people can't able to know about it and thus we think that the universe is borderless. And the light from the border of the universe will not reach us. We then see dark everywhere.

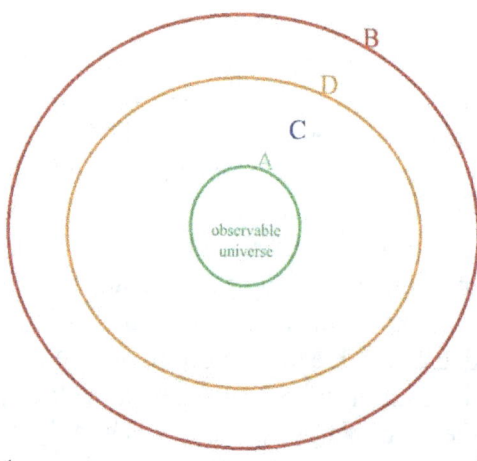

Figure. 14

In the above *figure.16*, it illustrates the universe and its boundary. The circle A is the boundary of observable universe. The circle B is the boundary of real universe. The circle D is the boundary line at which the light from circle B dissolves in space. The space C which is between the circles A&D is the space that our universe exists in. our universe should expand a lot to see the boundary of space.

The light cone illustrates the path of light and string cone illustrates the wave nature of light. From that extend by analyzing the *figure. 15* we can say that as time increases amplitude decreases and finally went to zero. Zero condition occurs at point P and at that point the light have swept completes maximum area.

By applying the boundary condition in *figure. 15* a new equation is formulated to calculate the space boundary. The equation provides the distance a light ray travels before it dies out. When considering the universe as whole the total intensity is to be calculated and that intensity determines the universal boundary.

$$D = \sqrt[4]{\frac{3C^3 Kx}{\pi\lambda^3}}$$

'D' is the total distance travelled by light before it completes it life journey. 'C' is the speed of light. 'K' is a constant. 'λ' is the wavelength of light.

Figure. 15 Scan the barcode for complete equation derivation.

The distance 'D' is called boundary distance and if we observe beyond this distance we can't able to see or experience the reality which ends in the boundary.

If we calculate the answer will go more than one septillion (a digit with 24 zeros).

12.1.4 Resultant

It is found that the reality that we claim as reality is not reality. It is a mixture of false realities. We all don't exist rightly as everything is false reality and mere imaginations. This also questions our existence in the sphere of reality and questions the difference between reality and dream as most times we got confused with realities and dreams in our daily life. As we don't exist, life also not exists, that means after life also didn't exists truly by science.

It is finally concluded as follows.

- ✓ There is nothing called reality exists.
- ✓ Everything maybe a mere dreamlike matter created by brain and our consciousness.
- ✓ There is a visual boundary for the universe.
- ✓ The light also dies (attenuates) after some time.

APPLICATION: 2

12.2 DNA CLOUD COMPUTING

Every human being has a "key code" in their DNA which helps or influences the behavior of the bio energetic waves which is emitted from the body of that living being. All living organisms (those have mental system and the ability to recollect) are storing their experiences and mental emotions in a cloud server and uses cloud computing technology in the nature. The key code in the DNA helps to enter into their stored data. We describe some impressive features of biological data storage by the nature, and speculate on approaches to research and development that could benefit the understanding about our memory.

12.2.1 OBJECTIVES OF THE STUDY

- ✓ To study about the memory storage systems in living beings especially humans.

- ✓ To study about different systems those are participating in the process of storing memory.

- ✓ To study about the bio energetic waves and its function in storing the memory process.

✓ To explain scientifically the Psychic and meditational methods in a more precise manner.

12.2.2 FROM THE PAST

When we dive into the helical structure of DNA through the mega project named Human Genome project (HGP). The project launched in 1990 and last for 13 long years and finally ended in 2003. The project got funded by many countries including US, Germany, China, Japan etc. It is considered as one of the greatest project ever conducted in the field of biology. The project aims to identify approximately 25000 genes in DNA. It also finds the played behind DNA, 3 billion chemical base pairs. The discovery is widely proclaimed as the cracking of god's code. The project give birth to new field of study called Bioinformatics.

In earlier days when the people are completely unaware of the god's code. They use similarities and differences to solve paternity issues. As this method is based on similarities and mainly females got the burden in their hands. Many loss their lives due to this unscientific method of finding parents. All the beliefs about birth and sex determination of a child got buried when HGP sequenced genetic codes and it got wide acceptance. New methods emerged for paternity determination such as DNA finger printing. Today DNA hold place in highly protected security devices and checks.

After 1950s many theories prophesized the time travel ability of DNA and they believe that we can recollect the lost memories of our ancient ancestors. Some researched in 1960s claimed that we have wormhole to our past experiences in our DNA.

Studies and researches are taking place to cure diseases from the genetic level and to create a disease free world in the near future. Genetic engineering also promising the dream world of new species and super humans. Many researches are carrying out to find the past live and to prove reincarnation based on genetic memory.

Films got produced taking these themes and last man on the line is movie in Tamil language titled "Eyam Arivu" starring Surya also deals with genetic memory and retrieving of this memory. The film also tells that the genetic memory retrieving help to get back even the memory that the person gained by his lifetime.

12.2.3 BLUEPRINT, MEMORY BANK, INNER
SPACE

The DNA within all living things is the blueprint for what each organism becomes, subject to the environmental influences that can also have significant effects.

For humans, recent discoveries about DNA are rapidly changing our views about the importance of this material. DNA may affect us much more significantly than we imagined. And, it may hold keys to further discoveries.

It has long been known that our physical appearance is determined by the combination of DNA from our mother and father. Now, researchers are confirming that certain diseases and disorders have direct links to our DNA. Our health may be programmed to some degree by our genetic history.

Our IQ and aptitudes, musical skills, athletic ability, even our psychological and emotional traits may be significantly affected by the DNA within us.

It has been demonstrated that experiences necessary for survival of a species are learned and that this knowledge is passed on to subsequent generations. In some cases this is mostly likely at least partially through DNA and the unconscious "instinct" that results. Even tiny and simple organisms learn crucial survival skills and pass these on.

For humans, with our relatively complex brain, feelings and memories, what other kinds of experiences might be saved in our DNA over the many thousands of years when our ancestors were born, lived and died? And, can they be accessed by us here and now?

12.2.4 WHAT WE KNOW AND DON'T KNOW

Scientific researchers are gradually uncovering the secrets of our DNA. They have identified the functions of and relationships between some of this material. Many genes remain a mystery and their purpose is unknown.

Sometimes, these mystery genes are called "junk DNA." According to some researchers, this may be an inaccurate label. Because the purpose and nature of this DNA material is not understood, it certainly does not mean it is useless junk.

As is often the case in scientific discovery, the more we know, the more we realize how little we know. Each question answered can raise many new questions.

For some, our human overconfidence and even arrogance can sometimes trick us into believing that we know all of the answers.

However, in the field of genetics research, there seems to be so much that is not known, that for an open-minded person, these kinds of theories about deep DNA memories cannot be ruled-out.

To conduct our own personal research and to find out for ourselves, maybe all we need to do is listen to our inner DNA.

Listen to the voices, feelings, sights and experiences of our ancestors. Their lives, joys and fears are within us. In that way, they are with us always.

> How can it possible to store a vast amount of data in a single DNA?

> Why all researchers are beating around the term DNA?

The answers are that the DNA is not able to store this vast amount of data which include the body system functionalizes, recipe to make a human and also the undiscovered assumptive existence of data about their ancestors also. We knew that even in this ultra-ultra-modern technology period, we do not have a storage device as small as DNA to store everything. The problem not lies in how the DNA can store but it actually lies in how much the DNA can store?

Consider when we went for a movie in a theatre. Even after some many years we don't forgot that movie. We also knew that a movie which is shown in a theatre must have a size above 200 GB depending on its clarity. After storing that movie in our memory we have lost 200 GB of space from our brain or DNA. Similarly as we go through every event that had taken place in our life consumes a memory about trillions and trillions of petabytes.

We know that the small microscopic DNA do not have the ability to store this vast amount of data. We also knew that genetic memory only retrieves our biological basic data and not details about our ancestral and the knowledge and experiences that they gained from their life.

Here we are disclosing the secrets of memory storage and rebuilding the structure without damaging the major foundational findings.

12.2.5 RESULTS

The 'junk DNA 'or the elements in DNA have a special key code. This key code influences all the neural mechanisms of the body and thereby stabilizes the energy which is emitted from the body. This emitted bio energy (aura) is consists of specific frequency which is common to that particular person or group or family.

The whole events that takes place is stored in the central controlling system of the universe in the form of energy which can be accessed. Each individual can access this data with the help of key code which is get transmitted from his or her body through their aura.

12.2.6 CLOUD COMPUTING AND HUMAN DNA

We have a whole bunch of storing techniques or devices today. Recently we have experience a new development of storing data in protein chains. We also have researches that are based on string data in our DNA.

Before going to the vast world of genetic memory I request you to get yourself free from the current dimension of viewing and transplant yourself to a new logical viewing.

12.2.7 CLOUD TECH

In the field of communication technology, we have a new comer i.e. cloud computing. This is same as what is happening inside our world between human beings. In cloud, we are accessing the data which is stored in a particular server from any part of the world. The server contains data about almost everything.

Similarly our memory is not stored inside us. The memories of human beings are stored in a vast storage center of the universe. Like we access the data from the cloud server, we can access the data's from the universal server.

12.2.8 THE SYSTEM

A system (here take a PC) is needed for accessing the data's from the cloud. Like that, we are accessing the universal cloud where our data's are stored from a system called human body.

12.2.9 THE LINK

We now know about the system which enables us to connect with the universal server called our body. But how this connection takes place? As like a computer, human brain just contain a RAM and a specific operating software called Mind by itself. The system is getting connected to the server by linking of aural energy.

12.2.10   SAFETY SECURE

Now we know about the system as well as the aural wave which act as an internet to the server. We know about the system hackers then how could it possible to keep the data's private and not mixed up?

Here is our DNA play its role. The DNA contains a specific password which is used to control and coordinate the wave frequency into uniqueness and thereby the wave from a particular body acts as a password for the particular data stored by him. A good hacker can hack this too.

12.2.11DATA ARRANGEMENT

The data's which are stored in the server is arranged as like a 'cone'. The bottom point 'P' in *figure.18* comprises of the very first folder which have the data of our first memory. I.e. childhood memories. As time passes and age increases more and more folders gets added with specific and nonspecific criteria.

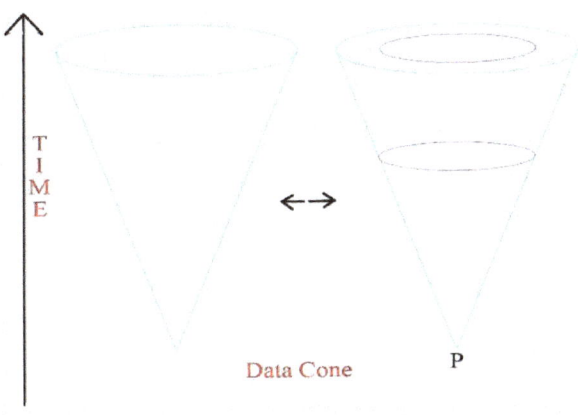

Figure. 16 Data storge cone illustration

The folders can be divided into two:

> Specific: These are those that have specific data's about an event or anything. These folders sometimes have more passwords to get connected by more than one.

> Non-specific: These are the folders that have some normal, unspecified data. These are those that are said to be stored in short term memory.

12.2.12 How this storage is created?

The mega storage system is just the cognitor wave action and its remaining effects. All the events are coded by cognitor waves and these waves get attenuated in pass of time. But there remains signatures of each event. These remains are termed as cognitor fossils. Because of these cognitor fossils we are able to know and feel certain past things when we visit certain places.

12.2.13 Mechanism

Each event constitute its own cognitor wave. This cognitor wave act as a cognitor memory bank which anyone can access. We simply copy down our needed details and create our own space of memory. Our brain just have memory like ram of the computer. This memory is contains data for the reflex actions and emergency measures. The universal data is exists even after ones death and can be access by anyone with the key wave. The data is transferred from person to person and person to their or others memory center using cognitor waves.

Figure. 17 Interconnection to server from different levels.

12.2.14   Resultants

✓ Each aural energy has its own specific

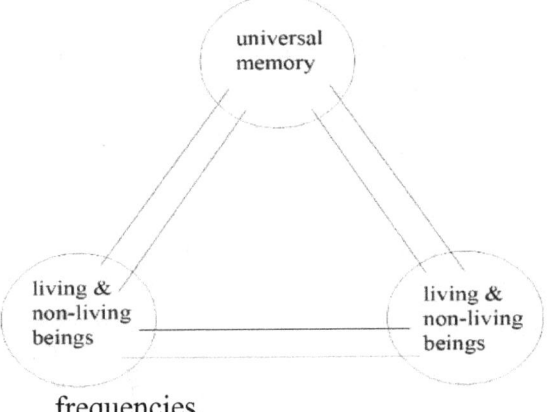

frequencies.

- ✓ The total memory is stored in the central server of the universe.

- ✓ The brain only has a functional memory and it is just like RAM in computers.

- ✓ Psychologists and psychic specialists are the hackers (make use of) of others frequency.

- ✓ The energy allows connection between brain to brain and brain to universal server.

APPLICATION: 3

12.3 When Sci-Fi came to reality!!

12.3.1 The dream for Elysium

In 2013 a sci-fi movie released named Elysium. The movie picturizes a world outside earth constructed by humans where there is no diseases and all. It is like heaven. The problem is only high-fi people reside there. All the others reside in earth which is in its extinction. The movie portraits a medical device. The device is capable of curing all diseases.

Now in reality we can construct such a devise by studying cognitor waves and cognitor fields in detail. As each component have its own cognitor wave, if we can able to recreate these waves then the body came to existence from nothing. So it will be a breakthrough in medical field. All diseases can be cured by studying and eliminating the cognitor wave which cases the disease by promoting and treating with an anti-cognitor to nullify the disease wave.

12.3.2 Manipulation technology in the movie Frequencies.

In 2015 a movie released which plays the story of frequencies and its power of manipulation. The in-depth study of cognitor waves help to make the device which is shown in the movie that is introduced as a 'true love finder'. All those manipulations can be down with cognitor waves. In the movie certain words are used to nullify the effect of cognitor waves and to adjust the situation. In reality anything can act as anti-cognitor in order to nullify the cognitor. Words re just one among the many.

13 Consequences.

As we are altering the system of cognitor by creating it in any aspect, we need to face its aftereffect's too. As everything is linked, when any part of system changes it affects the system and also the surrounding. The effect thus created may be small or big. But we have to face the aftereffects.

14 Appendix

The further study in the mentioned concept in the chapter "Realitiness of Reality" help us to reanalyze the use of light cones in the varied fields of study and it also help us to explore more about the realities around us. Further exploration in this field give more interesting results.

15   Proofs of cloud memory.

15.1   100th monkey effect

This is a phenomenon which describes how an idea or data is transferred from a group to another. The theory behind this phenomenon is proposed by Lawrence Blair and Lyall Watson in the late 1970's. The phenomena is claimed to be observed by Japanese scientists. They were observing a Japanese monkey Macaca fuscata for over 30 years in wild.

The scientists served sweet potatoes to the monkey and once in a while a little monkey came to learn itself that washing the potatoes help to get rid of sandy mud in it. The young monkeys teach this technique to heir elder ones. The twist came in the autumn season of 1958 some monkeys among Koshima group observed to wash potatoes. This transfer of idea from Macaca fuscata to Koshima monkey is the observed phenomenon.

15.1.1 Most of have experienced this. When you are in a crowd or maybe with some known or unknown person. Suddenly you can see that he talks or sings what you sing in your mind or what you wish to talk. The brains are connected each other. The data is transferred as thus we externalize our songs through other people. Just like a Bluetooth connection.

15.1.2 The existence of GPS system inside us. The brain by analysis the waves allow us to navigate ourselves.

15.1.3 Near my place there is forest area where the Government plant red oil palm over the area of 3646 hectares. Many domestic animals including goat, buffalo, cow, hen, duck etc. are raised here. None of the animals are tied or care taken by their owner. Animals are feely moving where ever they need to go. Even we humans find difficult to go back home if we're not a resident. All paths looks like and all leads to more deep inside. Interesting fact is that all the animals get back to their owner by evening 6 o clock. When the Asar adhan (Muslim prayer call of evening) is herd all the animals march to their respective place. Farmers don't lose their animal at any circumstances other than theft. These animals are well connected with the cognitor of the forest and we humans from outside not. So we find difficult to get out of the area.

15.1.4 Mind reading techniques (ESP).

15.2 Some extra notes

15.2.1 The cognitor waves can be used to identify, recreate everything.

15.2.2 It act as a universal barcode for everything.

15.2.3 Further study help to create a world without diseases. (the negative impact is there, but we can try to reduce it)

15.2.4 Study help to create life and put an end mark to the phenomenon death.

15.2.5 The characteristic features determine the wave from a body and waves that can accept by that body.

16 Acknowledgment

It's a great privilege to thank my teachers, friends, parents and all who knowingly and unknowingly support my experiments thereby providing data for my research work. My work have taken long five years for its completion. I am thanking Abhilash, who make me to question the connection between everything. Dr.Biju.S.Padmanabhan (Reiki specialist and psychic researcher) and Dr.V.George Mathew (retired Prof. of psychology Kerala University) for giving me access to people in varied fields of study.

I am extending my thanks to all my physics teachers especially Praveen Bose and Sameer who constantly supply inspiration to do more and my professor Ms.Aiswarya.T (Asst.Prof Applied Science; Physics, Vidya Academy of Science &Technology-Technical Campus, Kerala Technical University) who read and help me with corrections. Also Dr. Abdul Kalam (Professor in Physics. Iqbal College Peringamala, Kerala University) who provide me a national platform to present my paper about unified theory. I am also thanking Garrett Lisi (American theoretical physicist) who ask me to publish my paper as a book. Last but not the least I am thanking my friend Shameem and Soumya for their continuous push to make this book a reality.

17 Reference

No	Title	Author	Year	Source
1	The brief history of time	Stephan Hawking	1988	Book
2	The Theory of Everything (special edition)	Stephan Hawking	1996	Book
3	A lawyer presents the case for the afterlife	Victor Zammit		Online
4	ToE	Wikipedia		Online
5	The quest for theory of everything	Kitty Ferguson	July 1 1992	Online
6	Dead famous Albert Einstein and his inflatable universe	Dr.Mike Goldsmith	2001	Book
7	What is relativity?	L.D. Landau & G.B. Rumer	2003	Book

17.1 Other references of statements

Item	Source
Definitions and meanings	Google
Other data read	From all online sources

17.2 References of persons

Name	Designation
Praveen Bose	Physicist
Sameer	Physicist
Dr.V. George Mathew	Holigrative Psychologist
Dr. Biju S Padmanabhan	Reiki & psychic Specialist